Making and Selling Cosmetics Honeycomb Cleansing Cream
ISBN: 978-1-912271-56-6
Published by Northern Bee Books © 2020
Northern Bee Books, Scout Bottom Farm
Mytholmroyd, Hebden Bridge, HX7 5JS (UK)
www.northernbeebooks.co.uk
Tel: 01422 882751
Front cover image by Sara Robb. All other photographs by Rita Fitton.
To contact Dr Sara Robb: admin@drsararobb.info
Design by SiPat.co.uk

Making and Selling Cosmetics

Honeycomb Cleansing Cream

Dr Sara Robb

Contents

Foreword 6
Preface 7
Acknowledgements 8
Introduction 8
Cleansing Creams 9
The Role of Emulsifiers 9
Types of Emulsions 10
Cleansing Cream as an Alternative to Soap 11
Ingredients Used to Make Cleansing Cream 12
Functions of Ingredients in Cleansing Cream 13
Do I Need Specialist Equipment? 15
Making the Cleansing Cream 16
Packaging Cleansing Cream 20
Legal Requirements for Selling Cosmetics 21
Responsible Person and Product Information File 21
Cosmetic Product Notification Portal 23
Preservative Challenge Test 23
Cosmetic Product Safety Report 23
Labelling Honeycomb Cleansing Cream 24
Selling Cleansing Cream 25

Foreword

Dr Sara Robb was a huge influence in the creation and development of Bee Good skincare. It was attending one of her workshops at a British Beekeepers conference back in 2007 that inspired us to consider making lip balms and other cosmetic products from the honey, beeswax and propolis harvested from our bees in rural Hampshire. From her recipes and others researched from books going back to the 1700's we created an initial range that led to us winning "Best cosmetic product" five years in a row at the Hampshire Beekeepers show. Having launched as a professional business in 2013, Bee Good has become an award-winning, international skincare brand with many thousands of satisfied customers – all from a single, inspiring workshop from Dr Sara Robb!

Simon Cavill

Managing Director, Bee Good Ltd

Preface

Most of my life, I avoided putting oil on my skin; a rule I lived by. If you told me that I would someday sing the praises of cleansing cream, I never would have believed you. Come to think of it; if you had told me when I was a young girl in Iowa that I would one day live in London and teach beekeepers to make cosmetics, I would not have believed that either.

My interest in cleansing cream was sparked by reading old pharmacopoeias and cosmetic formularies. Each had a section on cleansing creams; oil and water emulsions used in the place of soap. I could not imagine how something oily smeared on your face could clean your skin and not leave you with a greasy complexion. After reading dozens of recipes, my curiosity got the better of me, and I made some cleansing cream.

To my great surprise, the cleansing cream left my skin clean and did not leave an oily residue. I expected a vicious flareup of acne, but the spots never arrived. In fact, after using the emulsion for a few days, my skin looked better than it had in years! Remember my rule not to put oil on my skin? What can I say? Rules were made to be broken!

Amazed by this revelation, I started wondering if cleansing cream would be helpful to pre-teen skin. My younger daughter is at the age where she is beginning to have breakouts. I whipped up a batch of cleansing cream, and I gave it to Meggy. Remarkably, Meggy's skin also benefitted from the cleansing cream, although we did adjust the formulation slightly.

The recipe in the first issue of Making and Selling Cosmetics is for Honeycomb Cleansing Cream. This cleansing cream is formulated with honey and beeswax and is suitable for all skin types. I hope you are as impressed as Meggy and I were.

Dr Sara Robb

Acknowledgements

Meggy Jayne deserves special recognition for her help with making the cleansing cream test recipes and letting me slather her with experimental formulations!

Photographer Rita Fitton took the pictures in this book, for which I am very grateful.

Finally, I would like to thank my publisher, Jeremy Burbidge, for encouraging me to write the Making and Selling Cosmetics series.

Introduction

Cosmetics containing bee products are gentle and naturally therapeutic. What's more, products made with beeswax and honey are moisturising and have a pleasant aroma. Golden soaps, delicious lip balms, and luxurious creams are just a few examples of cosmetics made with ingredients from the hive. The Making and Selling Cosmetics series will provide recipes for cosmetics incorporating bee products and discuss the steps required to sell these value-added products legally.

This edition of the Making and Selling Cosmetics series provides an easy-to-follow recipe for Honeycomb Cleansing Cream, a product that incorporates both beeswax and honey. Initially published in BBKA News, I wrote the Honeycomb Cleansing Cream recipe for my 2019 National Honey Show workshop. Honeycomb Cleansing Cream is a luxuriously thick emulsion that is fun to make and a delight to use.

Cleansing Creams

Cleansing creams contain a combination of oils, waxes and water in an emulsion. Applied liberally to the face, then simply wiped away, they simultaneously remove dirt and moisturize. Making emulsions can be challenging because the main ingredients are immiscible. What you need is a dependable formulation. With my recipe, you too can successfully make cleansing cream. Before we get started, I will provide a little useful background on emulsions.

The Role of Emulsifiers

We are all familiar with cosmetic emulsions, such as toothpaste, lotion and cream. But did you know emulsions are also in the kitchen? Indeed, you probably have attempted to make a culinary emulsion.

You can make a vinaigrette by whisking oil and vinegar together. The result of your labour is a thick and tasty emulsion. However, while you prepare your salad, the oil and vinegar part, in protest. Emulsion failure occurs because the oil and vinegar do not get along. The oil is hydrophobic (water-fearing), while the vinegar is lipophobic (oil-fearing), so they insist on going their separate ways.

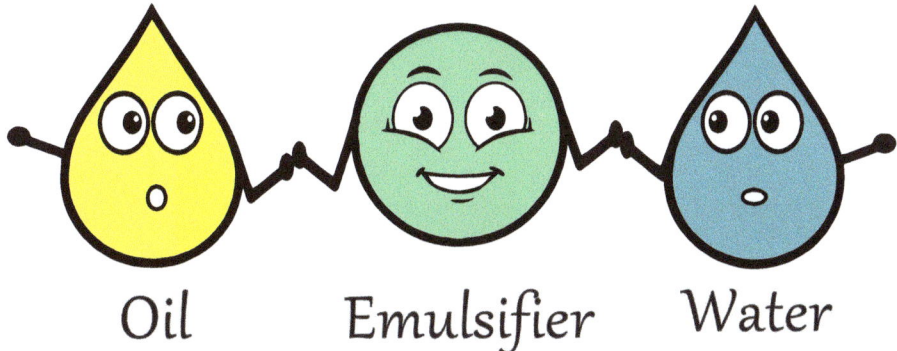

Figure 1. *Oil likes oil. Water likes water. But they do not like each other! Oil and water are reluctant molecules that do not want to interact, let alone mix! The emulsifier plays molecular matchmaker; taking the oil and water by the hand and coaxing them to form an emulsion.*

Oil and water can be stubborn and uncooperative ingredients, but a go-between can help. Emulsifier molecules contain a water-loving (hydrophilic) segment and an oil-loving (lipophilic) segment. The emulsifier acts as an intermediate that coaxes the reluctant water and oil molecules to interact (Figure 1). You could say the emulsifier is a molecular matchmaker, bringing the two timid phases together.

Most of us enjoy a glass of milk without thinking about its chemical organization. Milk is a perfect emulsion- it does not 'break' (separate). Milk proteins act as emulsifiers and keep the oil droplets suspended. Figure 2 shows a bottle of milk with the small spheres of fat coordinated in the aqueous background.

Figure 2. *Milk, is a perfect emulsion. Milk looks like a homogenous liquid to the naked eye, but milk is an emulsion. Small globules of fat are suspended in an aqueous background, resulting in its white appearance. The stability of milk is due to the presence of proteins that act as emulsifiers, holding the droplets of fat in place.*

Types of Emulsions

There are two types of emulsions, as shown in Figure 3. The first is an 'oil in water' (O/W) emulsion with the oil droplets dispersed in water- like milk. Conditioner, lotion and cleansing cream are examples of O/W emulsions. The second type of emulsion is 'water in oil' (W/O). In this case, water droplets are suspended in an oil background. Emulsions in the W/O configuration include butter and are often thicker than O/W emulsions. Because of the large amount of oil in W/O emulsions, these preparations can seem oilier than O/W emulsions.

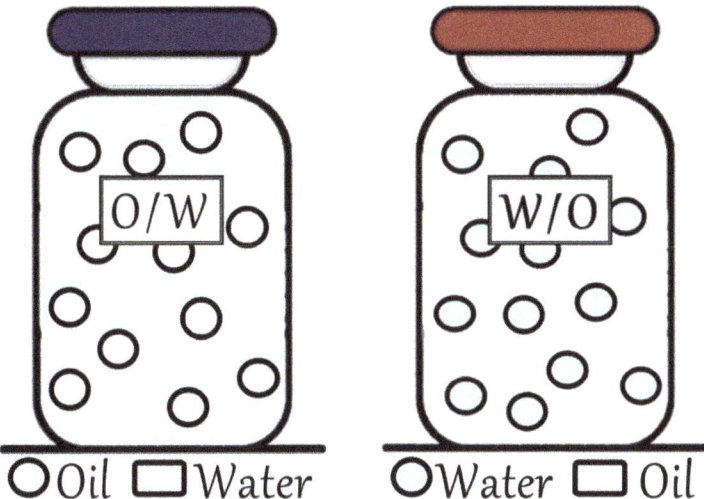

Figure 3. The two types of emulsion both contain oil and water, yet they differ in their configuration. On the left, an oil in water (O/W) emulsion with the oil droplets suspended in the aqueous background. On the right, a water in oil (W/O) emulsion with water drops coordinated in the oil background. Like milk, Honeycomb Cleansing Cream is an O/W emulsion.

While both types of emulsions are used to make cosmetics, it is somewhat easier to make a stable O/W formulation. Additionally, W/O emulsions break more easily. Many cosmetics take advantage of the stability of O/W emulsions, including Honeycomb Cleansing Cream.

Cleansing Cream as an Alternative to Soap

Soap is a cosmetic product frequently used to wash our skin. Soap molecules have a lipophilic end, which cleans oil-soluble impurities and a hydrophilic end, which removes water-soluble contaminants. Cleansing cream offers an alternative to soap. Made with oil and water cleansing cream also removes excess oil and water-soluble dirt.

Applied liberally to the face and simply wiped away, cleansing cream leaves skin clean and provides a protective barrier that does not feel greasy.

Ingredients Used to Make Cleansing Cream

Cleansing creams incorporate lipophobic and hydrophobic elements. These components are coaxed together by emulsifiers. The water or aqueous phase contains ingredients that do not want to interact with oil. In Honeycomb Cleansing Cream, water and glycerol are lipophobic. The oil or lipid phase in an emulsion contains the ingredients that do not want to combine with water. Mineral oil, stearic acid, lanolin, and beeswax comprise the hydrophobic ingredients in Honeycomb Cleansing Cream.

If we mixed the above ingredients, the oil and water would separate, leading to emulsion failure. We need a molecular matchmaker, an emulsifier, to entice the lipophobic and hydrophobic ingredients in the Honeycomb Cleansing Cream to form an emulsion.

While Figure 1 shows one emulsifier coordinating the oil and water molecules, cosmetic formulations often employ more than one emulsifier. Honeycomb Cleansing Cream incorporates polysorbate-80 and an emulsifying wax consisting of cetearyl alcohol and PEG-20 stearate- a total of three emulsifiers.

Functions of Ingredients in Cleansing Cream

Some ingredients used in the Honeycomb Cleansing Cream formulation were selected to make an emulsion. I chose additional components to give the final cosmetic a specific set of qualities. Table 1 summarises the functions attributed to each of the constituents used in Honeycomb Cleansing Cream. As you can see, most of the ingredients have multiple roles.

Water and mineral oil act as solvents in Honeycomb Cleansing Cream. These two ingredients facilitate the formation of the hydrophilic and lipophilic phases before they are brought together by the emulsifiers.

In Honeycomb Cleanser Cream, polysorbate 80, cetearyl alcohol and PEG-20 stearate are the emulsifiers that coax the oil and water phases into a stable configuration. Stearic acid further stabilizes the emulsion and adds cleansing action. Glycerin helps control the viscosity of the cream and is skin protecting.

Honey is emollient, humectant and moisturising, while beeswax provides skin conditioning and is film-forming. In combination, hive products add several ideal qualities to the final cosmetic.

Skin conditioning lanolin has the added benefit of reducing the skin's sebum production, making it an ideal ingredient to use in cosmetics for those with oily skin that is prone to breakouts.

The final ingredient added to Honeycomb Cleansing Cream is the preservative phenoxyethanol. Without a preservative, our cleansing cream would soon contain mould and could also be colonised by bacteria. A contaminated cream is not just unappealing; it is dangerous to use.

Cosmetic Ingredient	Function
Water	Solvent
Mineral Oil	Solvent, emollient, skin protecting
Stearic Acid	Cleansing, emulsion stabilising
Glycerin	Humectant, skin protecting, viscosity controlling
Lanolin	Emollient, skin conditioning
Polysorbate 80	Emulsifying
Cetearyl Alcohol	Emulsifying, emulsion stabilising
Beeswax	Emollient, skin conditioning, film-forming
PEG-20 Stearate	Emulsifying
Honey	Emollient, humectant, moisturising
Phenoxyethanol	Preservative

Table 1. The functions of cosmetic ingredients used to make Honeycomb Cleansing Cream. Many of the materials used to make cosmetics, including honey and beeswax, have multiple roles.

Do I Need Specialist Equipment?

It is a misconception that you need expensive, specialist equipment to make cosmetics. You can create fantastic products with the utensils you have at home. Most people will have all the equipment required to make Honeycomb Cleansing Cream.

Making cosmetics is very much like cooking; you need to measure, melt and mix. An accurate scale and weighing bowls are essential. To achieve the best results, ensure your scale measures in increments of one gram. We will use weight, rather than volume because it is a more precise way of measuring ingredients. A glass mixing bowl or a Pyrex jug is ideal for making cosmetics. Melting can be done on the hob or in a microwave, whichever you prefer. Spoons and a small hand whisk are sufficient to mix the ingredients in Honeycomb Cleansing Cream.

Making the Cleansing Cream

Honeycomb Cleansing Cream is made in three stages, as described below. Measure all the ingredients in grams (g). This recipe yields approximately 400 g of cleansing cream. You can scale-up as needed.

Recipe for Honeycomb Cleansing Cream
60 g Stearic Acid
8 g Beeswax
12 g Emulsifying Wax (Cetearyl Alcohol and PEG-20 Stearate)
12 g Polysorbate 80
40 g Glycerin
160 g Water
16 g Lanolin
92 g Mineral Oil
2 g Honey
2 g Phenoxyethanol

Stage 1

- Place the following ingredients in a heat-proof container: stearic acid, beeswax, emulsifying wax, polysorbate 80, glycerin and water. The contents of your bowl should resemble those shown in Figure 4.

Formulator's Tip
Use a bowl or jug with extra room, so your cleansing cream does not boil over when heated.

Figure 4. *The Stage 1 ingredients were all placed in a Pyrex jug. Note I have not mixed the contents at this point. After heating, the ingredients are mixed, then heated a second time.*

- Heat in the microwave or on a bain-marie to bring to a boil

- Mix the ingredients

- Return to heat until all solids have melted as shown in Figure 5

- Remove from heat and mix thoroughly

Formulator's Tip
You do not need to mix the ingredients before heating in Stage 1. Boiling begins the mixing process and saves you a messy spoon.

Figure 5. At the end of stage one, all the solids have melted. The bottom layer has a white appearance, indicating the emulsion has begun to form. However, there is a significant amount of oil floating on the top. The Honeycomb Cleansing Cream is not stable at this point.

Stage 2

- Weigh and add lanolin to the hot mixture (Figure 6)

- Stir until the lanolin dissolves

Formulator's Tip
Beeswax, stearic acid and the emulsifying wax used in this recipe all have melting points below the boiling point of water. The solids melt quickly in the boiling solution.

Figure 6. Add the lanolin to the Stage 1 ingredients. The warm liquid will melt the lanolin, so no additional heat is needed.

Stage 3

- Add mineral oil, honey, and preservative and stir to mix (Figure 7)

- Whisk until the emulsion is stable

- When Honeycomb Cleansing Cream has thickened, pour into containers

Formulator's Tip
To test the stability of the emulsion, stop stirring. If no oil floats to the top, the emulsion is stable. If you see small amounts of oil rise to the top, continue to whisk.

Figure 7. In Stage 3, add the mineral oil, honey and preservative to the warm mixture. Gentle whisking of the Honeycomb Cleansing Cream will create a stable emulsion. The cleansing cream will thicken as it cools.

Packaging Cleansing Cream

After making your Honeycomb Cleansing Cream, divide it into portions. You have your choice of plastic pots, glass jars or aluminium containers for your cream. If you would like to sell the cleansing cream, you will need to prepare uniform products. I decided to use 100 ml aluminium tins for my Honeycomb Cleansing Cream.

Legal Requirements for Selling Cosmetics

Three main conditions must be met to place a cosmetic on the market in the United Kingdom (UK). Currently, the regulations for selling cosmetics in the UK are the same as those of the European Union (EU). However, this may change after the Brexit transition period.*

The first is that there is a UK-based Responsible Person for the product. The Responsible Person must ensure the cosmetic product complies with UK/EU Cosmetics Regulation.

*The information provided here is correct at the time of publication.

Responsible Person and Product Information File

The second requirement is that the Responsible Person prepares a Product Information File (PIF) for the specific product (every product has its own PIF). Information contained in the PIF should include a description of the product and the Cosmetic Product Safety Report (CPSR), prepared by a suitably qualified assessor. An outline of the manufacturing process, including a statement about good manufacturing practice, will comprise part of the PIF. The Responsible Person should prepare a statement indicating that no animal testing has taken place to be included in the PIF.

The cosmetic's composition should be specified in the PIF, using the International Nomenclature of Cosmetic Ingredients (INCI). Include the formulation and a Technical Data Sheet (TDS) for each of the ingredients. See Table 2 for the INCI names and percentages of the ingredients used to make Honeycomb Cleansing Cream. If the product contains fragrance or essential oils, the International Fragrance Association (IFRA) certificates and allergen declarations are attached to the PIF. Results of stability testing and microbial analysis complete the PIF's documents.

Ingredient	INCI	Quantity (g)	Percentage
Water	Aqua	160.00	39.60%
Mineral Oil	Paraffinum Liquidum	92.00	22.77%
Stearic Acid	Stearic Acid	60.00	14.85%
Glycerin	Glycerin	40.00	9.90%
Lanolin	Lanolin	16.00	3.96%
Polysorbate 80	Polysorbate 80	12.00	2.97%
Cetearyl Alcohol	Cetearyl Alcohol	9.12	2.26%
Beeswax	Cera Alba	8.00	1.98%
PEG-20 Stearate	PEG-20 Stearate	2.88	0.71%
Honey	Mel	2.00	0.50%
Phenoxyethanol	Phenoxyethanol	2.00	0.50%

Table 2. *The information to include in the PIF for Honeycomb Cleansing Cream; The common name of each ingredient, followed by its INCI name, the quantity used in the recipe and the percentage of that ingredient in the final product.*

Cosmetic Product Notification Portal

The third requirement to comply with regulations is that the product is listed on the Cosmetic Product Notification Portal (CPNP) before being placed on the market. The CPNP is an online notification system which is free of charge. Each cosmetic requires a separate notification on the CPNP. Cosmetic product registration can be by an individual, the Responsible Person, or a company. Information such as the product name, images of the product and labels, manufacturer contact, product details and formulation is entered. After the Brexit transition period, there will be a UK based CPNP.

Preservative Challenge Test

Water-based cosmetics are required to undergo a Preservative Challenge Test (PCT). The PCT assesses the efficacy of the preservative used in a cosmetic product. A sample of the product is sent to a laboratory where it is 'challenged' with known quantities of micro-organisms. The product samples are monitored for nearly one month. During this time frame, if the preservative inhibits the growth of the micro-organisms, the cosmetic passes the test. If your aqueous product fails the PCT (micro-organisms grow in the product samples), you will need to adjust the formulation and submit the product again.

Cosmetic Product Safety Report

Following a successful PCT, a cosmetic safety assessor, such as myself, can provide you with a Cosmetic Product Safety Report (CPSR). You must also list the cleansing cream on the Cosmetic Product Notification Portal (CPNP) before it is placed on the market for sale.

Labelling Honeycomb Cleansing Cream

Labelling cleansing cream is relatively straight forward. The ingredients used to make Honeycomb Cleansing Cream will be listed on the cosmetic label by their International Nomenclature of Cosmetic Ingredients (INCI) names in order of descending quantity, as seen in Table 2.

Figure 8 shows the two labels prepared for Honeycomb Cleansing Cream. On the left, we see the name and intended use of the cosmetic product. The label on the right lists the ingredients in descending order by their INCI names. Next, is the batch number HCC0220, the Responsible Person (Bath Potions), and the company address. The nominal volume of each tin of Honeycomb Cleansing Cream is 100 ml. Finally, the cream should be used within 12 months of opening, as indicated by the open jar symbol.

Figure 8. *The labels prepared for Honeycomb Cleansing Cream meet the legal requirements. Left, a decorative one for the lid. On the right, the ingredient label for the bottom of the tin.*

Because Honeycomb Cleansing Cream does not contain essential or fragrance oil, there are no allergens to declare on the label. However, don't be tempted to claim your product is "fragrance-free" or "allergen-free." Since July 2019, making "free-from" claims is against the law. According to legislation (EU) 655/2013. It is also illegal to make statements such as "paraben-free" or "sulfate-free" because these claims disparage other products for containing ingredients that are legal to use.

Selling Cleansing Cream

Follow the recipe in this article, and you can make Honeycomb Cleansing Cream. This emollient emulsion, containing honey and beeswax (Figure 9), is an ideal cosmetic to add to any beauty routine. You are free to sell my recipes, provided you follow the legal steps to obtain a CPSR. Please contact me at Verdigris Cosmetic Solutions if you are interested in purchasing a CPSR or would like any other advice. Complete the steps above, including the challenge test, and your Honeycomb Cleansing Cream is ready to sell. Why not include this cleansing cream in your range?

Figure 9. Honeycomb Cleansing Cream, a luxurious W/O emulsion, packaged in a 100 ml tin, labelled and ready to sell at the market.

References

Robb SJ. Dr Sara's Honey Potions. Northern Bee Books, 2009.

Robb SJ. Beauty and the Bees. Northern Bee Books, 2012.

Bullen A. Selling Hive Product Cosmetics and the Law. BBKA News; September 2018, p309–310.

Robb SJ. Making and Selling Cosmetics: Emulsions. BBKA News; January 2020, p25-28.

European Commission Cosmetic Ingredient Database:
https://ec.europa.eu/growth/sectors/cosmetics/cosing_en

(EU) 655/2013 Regulation Common Criteria of Claims:
https://eur-lex.europa.eu/eli/reg/2013/655/oj

Index

A
Aqua (see Water) 22, 24

B
Batch Number 24
Beeswax 6, 7,8, 12, 13, 14, 16, 17, 18, 22, 25

C
Cera Alba (see Beeswax) 22, 24
Cetearyl Alcohol 11, 13, 14, 16, 22, 24
Cleansing Cream 7, 9, 11, 12, 13
Cosmetic Allergens 23
Cosmetic Product Notification Portal (CPNP) 23
Cosmetic Product Safety Report (CPSR) 21, 23, 24

E
Emulsifiers 9, 10, 12, 13
Emulsifying Wax 12, 16, 17, 18
Emulsions 7, 8, 9, 10, 11
Equipment 15

F
Formulator's Tips 17, 18, 19
Free-From Claims 24
Functions of Ingredients 13, 14

G
Glycerin 13, 14, 16, 17, 22, 24

H
Honey 6, 7, 8, 13, 14, 16, 19, 20, 22
Hydrophilic 10, 11, 13
Hydrophobic 9, 12

I
Ingredients in Honeycomb Cleansing Cream 14, 16
International Fragrance Association (IFRA) 21
International Nomenclature of Cosmetic Ingredients (INCI) 21, 22, 24

L

Labelling Honeycomb Cleansing Cream 24
Lanolin 12, 13, 14, 16, 18, 19, 22, 24
Legal Requirements for Selling Cosmetics 21, 24, 25
Lipophobic 9, 12
Lipophilic 10, 11, 13

M

Making Honeycomb Cleansing Cream 17, 18, 19
Mel (see Honey) 22, 24
Milk 10, 11
Mineral Oil 12, 13, 14, 16, 20, 22

N

Nominal Volume 24

O

Oil 7, 9, 10, 11, 12, 13, 14, 18, 19, 20, 22, 24

P

Packaging Honeycomb Cleansing Cream 20
Paraffinum Liquidum (see Mineral Oil) 22, 24
PEG-20 Stearate 12, 13, 14, 16, 22, 24
Phenoxyethanol 13, 14, 16, 22, 24
Polysorbate 80 .... 12, 13, 14, 16, 17, 22, 24
Preservative Challenge Test (PCT) 23

R

Recipe for Honeycomb Cleansing Cream 16
Responsible Person (RP) 21, 23, 24
Product Information File (PIF) 21

S

Selling Honeycomb Cleansing Cream 21, 24, 25
Soap 7, 8
Stearic Acid 12, 13, 14, 16, 17, 18, 22, 24

T

Technical Data Sheet (TDS) 21

U

Use Within 24

W
Water 7, 9, 10, 11, 12, 13, 14, 15, 16, 17, 22, 23

www.ingramcontent.com/pod-product-compliance
Lightning Source LLC
Chambersburg PA
CBHW041632040426
42446CB00022B/3484